Correlations between the Physical and Social Sciences

Valentine J. Belfiglio

UNIVERSITY PRESS OF AMERICA,® INC.
Lanham • Boulder • New York • Toronto • Plymouth, UK

Copyright © 2012 by
University Press of America,® Inc.
4501 Forbes Boulevard
Suite 200
Lanham, Maryland 20706
UPA Acquisitions Department (301) 459-3366

Estover Road
Plymouth PL6 7PY
United Kingdom

Library of Congress Control Number: 2011934408
ISBN: 978-0-7618-5589-7 (paperback : alk. paper)
eISBN: 978-0-7618-5590-3

To my wife Ellie, novelist and poetess extraordinaire, for always being Ellie.

Contents

Acknowledgment

I am grateful for the invaluable technical assistance given me by my friend and Graduate Teaching Assistant, Karen Webb, without whose help this monograph could not have been written.

Introduction

With the advent of the twenty-first century, the study of the social sciences had passed through four major stages, which Dougherty (2001) termed utopian; realist; behavioral; and postmodern, postpositivist. Most social scientists make normative judgments during their research, and many combine methodological approaches.[1]

The utopian approach features an imaginative creation of ideal states and social systems. Plato's *Republic* is an example.[2] The realist approach emphasizes the concept of national interests defined in terms of power. Power consists of anything that establishes and maintains control over the minds and actions of people. Realism conflicts with idealism, which advocates that foreign and domestic policies based on moral principles, are more effective than realism, because it promotes unity and cooperation rather than competition and conflict. Niccoli Machiavelli's *The Prince* is an example of the realist approach.[3]

Behavioralism is an approach that employs scientific methods and perspectives to the study of the social sciences. Behavioralists utilize statistics as a means of assembling, classifying, tabulating and analyzing phenomena. Jaral B. Manheim's (1990) *Empirical Political Analysis* is an example of the behavioralist approach.[4]

The postmodernist, postbehavioral, postpositivist approaches to the study of the social scientists have an obligation to become more relevant and concerned with values and to use their special knowledge to improve society. This has led to conflicting value judgments and standards characterized by diversity that includes paradigmatic disagreements and debates even about the ability to produce cumulative theories for the social sciences. Yosef Lapid's (1989) *Prospects of International Theory in a Post-Positivist Era*, is an example of the postmodernist, post behavioral, postpositivist approach.[5] The interparadigmatic debate has left social scientists without a center of gravity from which to advance systematic, harmonized, reproducible studies of people living together in groups, as families, clans, tribes, nations and states. Many postmodernists might relate to the dilemma expressed in the opening lines of Dante's *Inferno*:

> In the midway of this our mortal life,
> I found me in a gloomy wood, astray
> Gone from the path direct:[6]

What is to be done? The thesis of this monograph is that societies in general, are governed by objective laws that have their roots in human nature. The task of the social scientist is to discover and explore these laws. Empiricists were on the right track in their use of numerical data to classify and interpret events, but their studies did not go far enough. Null hypotheses and alternative rival hypotheses developed by social scientists must be eclectically correlated to mathematical formulae or the laws of physics in order to advance non-speculative, unbiased knowledge.

A preliminary assumption of this monograph is that there is a unity of all forms of knowledge, grounded in nature. Mathematicians and physicists are artists who use nature as their model to provide orderly and unifying interpretations. Social scientists must follow their example. The etymology of the word "science" comes from the Latin *scientia*, which means knowledge. Thoughtful scientists and philosophers have recognized the correlation of academic disciplines, and the potential unity of science and the humanities. Rene Descartes pointed out "the bond between philosophy or science and society in the common enterprise of 'the mastery of nature.'"[7] Edmund Husserl asserted "there is a need for a unified and unifying, comprehensive and all-clarifying knowledge."[8]

Samuel Coleridge had in mind "nothing less than the Platonic ideal of absolute universality."[9] Coleridge wrote: "in order to derive pleasure from the occupation of the mind, the principle of unity must always be present."[10] Albert Einstein concluded that "all knowledge about things is exclusively a working-over of the raw material furnished by the senses."[11] Therefore, the task of the physical scientist, social scientist, or one who attempts to correlate the physical and social scientists is the same: "methodological thinking directed toward finding regulative connections between our sensual experiences."[12]

Ludwig von Bertalanffy (1954) notes that "the physicist, the biologist, the psychologist and the social scientist are, so to speak, encapsulated in a private universe, and it is difficult to get a word from one cocoon to the other."[13] There have been attempts to unite segregated academic disciplines. Examples are: biological anthropology, social psychology, and political ecology. This effort to integrate academic disciplines is laudable but unsuccessful in establishing non-speculative, unbiased, eclectic knowledge.

The following four chapters contain case studies which attempt to correlate null or alternative rival hypotheses with mathematical formulae or the laws of physics, in order to advance knowledge about human nature. The first case study relates marital assimilation of minority groups into dominate core cultures, with Graham's Law for the diffusion of gases. The second case study relates the mutual hostility of political leaders, with the Mirror Equation employed in basic geometric optics. The third case study relates the duration of major American military conflicts to the formulae for empirical and subjective probabilities. The fourth case study relates the radioactive decay formula for radioactive substances to the rate of decline of seven extinct empires.

The four case studies in this monograph do not provide definitive answers. They are plausibility probes[14] designed to determine whether or not a multidisciplinary approach involving the physical and social sciences is a worthwhile endeavor. Human beings are vastly more complex than the properties, changes and interactions of matter and energy; or the quantities, magnitudes, forms, relationships and attributes of numbers and symbols. The laws of physics and science of mathematics can serve as a model only for the formation of hypotheses about human behavior.

Prior to a major battle between an ancient, Roman legion and an enemy army, the commander of the legion (legatus) ordered soldiers (ferentarii) to engage and "provoke the enemy, going in front of the line."[15] The ferentarii gained tactical information about the location, capabilities, and possible intentions of the other army. They then "returned to their own men and took up positions behind them."[16] Casualty rates among these advanced units were usually high, but they often prevented the main body of the legion from falling into a trap.

I am attempting to be among the ferentarii of scholars who probe potential relationships between the physical and social sciences, I might be successful in providing information and incentive for a legion of supporters. I might be unsuccessful and "suffer the slings and arrows of outrageous fortune."[17] In either event, I am content that I have undertaken a novel, multidisciplinary approach to the study of the social sciences.

NOTES

1. James E. Dougherty (et al.), *Contending Theories of International Relations* (New York: Longman, 2001), pp. 616, 653.

2. Jack C. Plano, *The American Political Science Dictionary* (New York: Harcourt Brace Jovanovich, 1993), pp. 26-27.

3. Jack C. Plano, *Political Science Dictionary* (Hinsdale, Illinois: The Dryden Press, 1973), p. 317; Hans J. Morganthau, *Politics Among Nations* (New York: Alfred A. Knopf, 1973), pp. 5, 9.

4. Plano, *Political Science Dictionary, op. cit.*, p. 28; Morris Kline, *Mathematics: A Cultural Approach* (Reading, Mass.: Addison-Wesley, 1963), p. 613.

5. Dougherty, *Contending Theories of International Relations, op. cit.,* p. 616; J.F. Lyotard, T*he Postmodern Condition: A Report on Knowledge* (Minneapolis: University of Minnesota Press, 1984), pp. 14-36.

6. Dante Alighieri, *The Divine comedy*, Translated by Henry F. Cary, *The Harvard Classics*, Vol. 20 (New York: P.F. Collier & Son company, 1909), p. 5. Dante, *Inferno*, Canto 1: 1-3.

7. Leo Strause, *History of Political Philosophy* (Chicago: University of Chicago Press, 1987), p. 435.

8. *Ibid.*, p. 879.

9. Walter Jackson Bate, *Coleridge* (London: Weidenfeld, and Nicolson, 1968), p. 149.

10. Samuel Taylor Coleridge, *On Poesy or Art*, Edited by Charles W. Eliot, *The Harvard Classics*, Vol. 27 *English Essays* (New York: P.F. Collier & Sons, 1910), p. 277.

11. Albert Einstein, *Ideas and Opinions* (New York: Crown Publishers, 1954), p. 21.

12. *Ibid.*, p. 50.

13. Ludwig von Bertalanffy, "General Systems Theory," in J. David Singer (Ed.), *Human Behavior and International Politics: Contributions from the Social-Psychological Sciences* (Chicago: Rand McNally, 1965), p. 21.

14. Harry Eckstein, "Case Study and theory in Political Science" in Fred I. Greenstein and Nelson W. Polsby (Eds.), *The Handbook of Political Science*, Vol. 7, *Strategies of Inquiry* (Reading, Mass.: Addison-Wesley, 1975), pp. 92-123.

15. Publius Flavius Vegetius Renatus, *Epitoma Rei Militaris*, Translated by Valentine J. Belfiglio, Libellus II, Vegetius, II: xvii.

16. *Ibid.*, Libellus II: xvii.

17. William Shakespeare, *Hamlet*, in W.G. Clark (Ed.), *The Complete Works of William Shakespeare*, Vol. 2 (New York: Nelson Doubleday, 1853), p. 613. Shakespeare, *Hamlet*, III.1.56.

Chapter One

Exogamous and Endogamous Marriages among Italians and Mexicans in Dallas County, Texas (2000)

. . . Nature has ordained that we (men) can't live easily with women or without them,"

—Metellus Macedonius, Speech to the Roman Senate (131 B.C.)

INTRODUCTION

The concepts of acculturation and assimilation, though controversial, remain popular among scholars and students of racial and ethnic relations. For the purposes of this study, "acculturation" means the process of adopting foreign cultural patterns through contact with another people. "Assimilation" means the absorption of an ethnic population, often the descendants of immigrants, into the dominant society accompanied by the partial or total loss of a separate identity.[1] Numerous books and articles focus on racial and ethnic relations in the United States. Two major themes emerge from these studies: assimilation theories, and power-conflict theories. Assimilation theories emphasize conformity to the dominant Anglo-Saxon-Protestant culture and society, or cultural pluralism outcomes.[2] Power-conflict theories place greater emphasis on economic stratification and power issues, as they accept substantial inequality and stratification.[3] This paper utilizes the assimilation theory because that theory is more amenable to its purpose: endogamy and exogamy among Italians and Mexicans living in Dallas County, Texas.

PURPOSE AND METHODOLOGY

Assimilation analysts emphasize different dimensions of intergroup adaptations, such as acculturation and marital assimilation. Milton Gordon (1964) identifies seven processes of the adaptation of an ethnic group to a core society. These processes include:

1. Cultural assimilation: change of the cultural patterns to those of the core society.
2. Structural assimilation: penetration of cliques and associations of the core society at the primary-group level.
3. Marital assimilation: significant intermarriage.
4. Identification: development of a sense of identity linked to the core society.
5. Attitude-receptional assimilation: absence of prejudice and stereotyping.
6. Behavior-receptional assimilation: absence of intentional discrimination.
7. Civic assimilation: absence of value and power conflict.[4]

This study will focus on the marital assimilation process of Gordon's adaptation theory. It will compare and contrast endogamy and exogamy[5] among Italians and Mexicans living in Dallas County, Texas in 2000. The thesis of this paper is that exogamy among Italian and Mexican Texans is dependent upon the population concentration of the ethnic group in relation to the population concentration of the core group. The greater the relative concentration of an ethnic group to the concentration of the core group, the less likely exogamy will take place.

This study will employ the Chi-Square test, and a formula extrapolated from Graham's Law of Diffusion in an effort to prove the thesis statement. The Chi-Square test is one of the most useful and versatile of all statistical tests[6] and will be applied to determine whether or not marriage preferences among Italian and Mexican Texans is associated with their ethnicity.

No formula has been developed by social scientists to measure the rate of marital assimilation of ethnic groups. Therefore, as the field of biological anthropology is grounded in the natural sciences rather than social studies, this study will turn to the natural science by adapting Graham's Law of Diffusion in an effort to determine the marital assimilation rate of ethnic groups.

Graham's Law, developed by the chemist, Thomas Graham (1805-1869), states that the comparative rates of diffusion of two gasses is inversely proportional to the square root of their densities.[7] Human beings are far more complex than inorganic gases, or even other living species. Marital assimilation and the diffusion of two gases are quite dissimilar. From a knowledge of the diffusion of gases, we cannot directly infer anything about exogamy. In some respects, though, they may be analogous, and a comparison of basic similarities and differences can help scholars avoid oversimplified single-factor explanations. Diffusionism is fundamental to anthropological inquiry in explaining similarities in the habits and beliefs of people living in under-developed parts of the world. Adapting Graham's Law to marital assimilation would state: The comparative rates of marital assimilation of two cultures is inversely proportional to the square root of their population densities.

This study employs a bivariate analysis by employing statistics summarizing the relationship between two variables. As such it is useful as a prototype for a more in-depth multivariate analysis which is beyond the scope of this paper. A multivariate study could include variables such as income, gender, religion, political orientation, generational differences, educational levels, etc.

THE DALLAS COUNTY EXPERIMENT

Studies by Alba (1985) and Feagin (1989) indicate that Italian Americans have an exogamous marriage rate of nearly 80 percent.[8] Other studies by Murguia (1982) and Feagin (1989) demonstrate that Mexican Americans have an exogamous marriage rate of 24 percent.[9] Between 2001 and 2003, Belfiglio interviewed scores of Italians and Mexicans in Dallas County, Texas in an effort to compare and contrast the endogamous and exogamous marriage rates of these two ethnic groups.

He conducted a countywide rather than a statewide study to minimize variables associated with political, economic, social, geographic, historical and cultural factors. Dallas County offered several advantages. First, Dallas County is his home, and this fact facilitated the interview process. As a member of the Italian Club of Dallas, it was easy for him to approach members of the organization.[10] As a part-time pharmacist at a pharmacy located in the Hispanic community, he was in touch with scores of families of Mexican descent.[11] Second, closeness to the Office of the County Clerk, allowed him access to Certificates of Marriage, which are located in that office. Third, the percentage of Hispanic population (29.87%) and Italian population (2.1%) are close to the percentage of Italians and Mexican living in Texas. To control for the influence of religion on the dependent variable, only marriages between Roman Catholics were taken into account. The reason for this selection is that 88% of the Italian Texans and 96% of the Hispanic Texans who married during the period under study were Roman Catholics.

Table 1.1 gives the number and percentages of Italians and Mexicans living in Texas in 2000. Tables 1.2 and 1.3 show the number of endogamous and exogamous marriages of Mexicans and Italians living in Dallas County in 2000. The Chi-Square tests found in Tables 1.4 and 1.5 clearly demonstrate that marriage preferences among Mexican Texans is associated with ethnicity, and that marriage preferences among Italian Texans is not associated with ethnicity. Table 1.6, based upon an extrapolation of Graham's Law of Diffusion,[12] indicates that the marital assimilation rate for Mexican Texans is 3.8 times greater than the marital assimilation rate for Italian Texans.

CONCLUSION

This study indicates that a majority of Mexican Texans preferred to marry within their ethnic group in 2000. This was not the case for most Italian Texans. Since most of the Italians studied in this project are descendants of immigrants who migrated to America from 1880-1922, they were already assimilated into the core group. Many of the Mexican Texans interviewed were more recent arrivals to the United States, and were not as well assimilated.

Table 1.5 predicts a marital assimilation rate of 0.547 for Mexican Texans. Future studies, based on the U.S. Census figures for 2010 will support or refute the findings of this report, and the utility of the extrapolated formula as a predictor of future marital assimilation rates. The thesis of the paper has changed as a result of the findings: The greater the relative concentration of an ethnic group to the concentration of the core group, the more likely exogamy will take place.

It is now appropriate to make normative statements which expound a subjective, preferential point of view in contrast to an empirical statement that seeks to describe what actually exists. I believe that social unity is better than social disunity and that ethnic tolerance and respect is superior to ethnocentrism. I also believe that pride in one's citizenship is more important than pride in one's ethnic group. Ethnocentrism can lead to a belief that one's own ethnic group and culture is superior to all other ethnic groups. Ethnic tolerance and respect can lead to an acceptance of other ethnic groups as equal, valuable human beings. The ultimate fruition of ethnic tolerance and respect is intermarriage among all ethnic groups. This position is compatible with the constructivist approach to an understanding of ethnicity. Other scholars could choose to examine Italian and Mexican marriages from the primordial or instrumentalist approach to the study of ethnicity, which was beyond the scope of this study. Generally, primordial theories assert that ethnic identification is based on deep, primordial attachments to a group or culture; instrumentalist approaches threat ethnicity as a political instrument exploited by leaders and others in a pragmatic pursuit of their own interests.

The intervening variables of generational differences, and social stratum should be included in future, more in-depth studies of endogamous and exogamous marriages. By social stratum is meant, families or individuals located at the same level on some aspect of social rank, such as income, wealth, power, or prestige.

Table 1.1. Italian and Mexican Residents in Texas (2000)

Ethnic Group	Number	Percent of Population
Italians	363,354	1.7%
Mexican	6,669,666	31.99%
Others	13,818,800	66.31%
TOTAL	20,851,820	100.00%

Sources: U.S., Bureau of the Census, Department of Commerce, 2000. Statistical Abstract of the United States, Census Bureau

Table 1.2. Marriage among Mexican Texans in Dallas County, Texas (2000)

Marriage Preference	Mexican Texans	Other Texans
Endogamy	4,415	1,159
Exogamy	1,035	10,428
TOTAL	5,450	11,587

Sources: U.S., Bureau of the Census, Department of Commerce, 2000; Statistical Abstract of the United States, Census Bureau; Texas vital Statistics (2000), of the Texas department of health. Additional information was obtained by interviewing, in a systematic and comprehensive way, scores of residents in Dallas county. The interviews occurred from 2001-2003.

Table 1.3. Marriage among Italian Texans in Dallas County, Texas (2000)

Marriage Preference	Italian Texans	Other Texans
Endogamy	61	5,513
Exogamy	229	11,234
TOTAL	290	16,747

Sources: U.S., Bureau of the Census, Department of Commerce, 2000; Statistical Abstract of the United States, Census Bureau; Texas vital Statistics (2000), of the Texas department of health. Additional information was obtained by interviewing, in a systematic and comprehensive way, scores of residents in Dallas county. The interviews occurred from 2001-2003.

Table 1.4. Mexican Texans Married in Dallas County, Texas (2000) (in thousands)

Cell	Observed	Expected	Difference	D^2	D^2/E
Mexican Endogamy	4.42	1.78	2.63	6.9	3.9
Other Endogamy	1.16	3.79	−2.63	6.9	1.8
Mexican Exogamy	1.04	3.67	−2.63	6.9	1.9
Other Exogamy	10.43	7.80	2.63	6.9	0.66
Degrees of Freedom: 1				$x^2=$	8.26

Ho: Marriage preference among Mexican Texans is not associated with ethnicity.
Alternative: Marriage preference among Mexican Texans is associated with ethnicity.
Significance level: 0.5
Statistic: x^2
Value of x^2 to reject Ho: 3.841

Sources: Table 1.2

Table 1.5. Italian Texans Married in Dallas County, Texas (2000) (in tenths)

Cell	Observed	Expected	Difference	D^2	D^2/E
Italian Endogamy	6.1	9.5	−3.4	11.56	1.22
Other Endogamy	551.3	547.9	3.4	11.56	0.02
Italian Exogamy	22.9	19.5	3.4	11.56	0.59
Other Exogamy	1,123.4	1,126.8	−3.4	11.56	0.10
Degrees of Freedom: 1				$x^2=$	1.93

Ho: Marriage preference among Italian Texans is not associated with ethnicity.
Alternative: Marriage preference among Italian Texans is associated with ethnicity.
Significance level: 0.5
Statistic: x^2
Value of x^2 to reject Ho: 3.841

Sources: Table 1.3

Table 1.6. Population of Dallas County, Texas (2000) (in ten thousands)

Ethnic Group	Number	Percentage
Italians	4.66	(2.1%)
Mexicans	66.28	(29.87%)
Other	150.95	(68.03%)
TOTAL	221.89	(100.00%)

Sources: Texas, Dallas County, Office of the County Clerk, Certificates of Marriage (2000). Additional information was obtained by interviewing, in a systematic and comprehensive way, scores of residents in Dallas county. The interviews occurred from 2001-2003.

$$MAR = \frac{\sqrt{D^2}}{\sqrt{D^1}}$$

Where: MAR = the marital assimilation rate.
D2 = population density of the ethnic group.
D1 = population density of the entire county.
MAR for Italians = 0.145   MAR for Mexicans = 0.547.

NOTES

1. Leonard Broom (et al.) Sociology (Belmont, California: Wadsworth, 1993), p. 348.

2. Robert E. Park, Race and Culture (Glencoe, Illinois: Free Press, 1950) p. 150.

3. Mario Barrea, Race and Class in the Southwest (Notre Dame, Indiana: University of Notre Dame Press, 1979), pp. 214-217.

4. Milton M. Gordon, Assimilation in American Life (New York: Oxford University Press, 1964), pp. 71-73.

5. Exogamous marriage-outside a given group or category. Endogamous marriage-within a given group or category. Allan Barnard, (et al.), encyclopedia of social and Cultural Anthropology (London: Routledge, 2001), pp. 603, 605.

6. Oliver Benson, Political Science Laboratory (Columbus, Ohio: Charles E. Merrill, 1969), p. 154.

7. Paul Peter Urone, college Physics (Pacific Grove, California: Brooks/Cole, 2001), pp. 287-290.

8. Richard Alba, Italian Americans (Englewood Cliff, New Jersey: Prentice-Hall, 1985), pp. 159-162; Joe E. Feagin, Racial and Ethnic Relations (Englewood Cliff, New Jersey: Prentice-Hall, 1989), p. 137.

9. Edward Murguia, Chicano Intermarriage: A Theoretical and Empirical Study (San Antonio, Texas: Trinity University Press, 1982), pp. 45-51; Feagin, Ibid., pp. 278-279.

10. Italian Club of Dallas, 14865 Inwood Road, Dallas, Texas.

11. West Dallas Pharmacy, 1215 Singleton Blvd., Dallas, Texas.

12. For Graham's Law of Diffusion, see the Appendix.

Chapter Two

The Mirror Image:
A Correlation between
the Physical and Social Sciences

Mirror Image-"an image in reverse; reflection,"

—Webster's New World Dictionary.

INTRODUCTION

This article will quantify and classify the intensity of hostility between the Administrations of Presidents George W. Bush of the United States, and Mahmoud Ahmandinejad of Iran, between 2005-2008. The key question this work seeks to answer is: Does the Mirror Equation employed in basic geometric optics correlate with the concept of the mirror image used in international relations theory? The salient information in this study is speeches by Bush and Ahmandinejad which refer to each other's government. The main inference of this work is that an interdisciplinary approach can challenge the common narrowness of ingrained, discipline-bound ways of thinking that produce occlusion as well as insight.

The key concept needed to understand this article is: If the knowable universe is a unity, discipline is a loss as well as a gain, and interdisciplinary, trans-disciplinary or multidisciplinary foci can effectively cross disciplinary boundaries to enhance understanding. Common models employed in physics and international relations theory may need to be balanced with and measured against the broader perspective of interdisciplinary, in order to better understand the concept of mirror image. If we take this line of reasoning seriously, the implication is a reconstructive, holistic connection between and among the physical and social sciences, and an escape from the occlusions of narrow, methodological approaches to knowledge. If we fail to take this line of reasoning seriously, the implication is at times, to leave distinctive disciplinary modes stranded, and separated from their natural roots and social ends.

THE PATHOGENESIS OF HOSTILITY

The genesis of hostility between Iranian and American leaders is complex and difficult to determine with definitiveness. A salient event occurred on November 4, 1979 when militant Iranian students occupied the American embassy in Tehran, with the support of Ayatollah Khomeini. In response, U.S. President Jimmy Carter severed diplomatic relations with Iran on April 7, 1980. Then, the United States government suspended most trade with Iran and imposed economic sanctions with the Islamic Republic.[1] However, the seizure of the American embassy and the American response were not the cause of hostility, but rather a result of pre-existent hostility.

Dominant theories of hostility, aggression and war are materialist, and involve competition over scarce or limited vital resources, such as land, food and trade. The specific scarce resource in the Middle East and Near East is petroleum.[2] Another explanation for hostility, aggression and war emphasizes differences in cultural values and beliefs.

Huntington (1996) asserts that culture and civilization will define future armed conflicts. According to him, "war. . . could come about from the escalation of a fault line between groups from different civilizations, most likely involving Muslims on one side and non-Muslims on the other. . . "the United States . . . defines its global role as the leader of Western Civilization.[3] Thus, the hostility and conflict between Iran and the United states can be viewed as a clash between Islamic and Western civilizations. If Huntington is partially correct, then neither Islam nor Christianity is pertinent to the clash of Western and Persian civilizations. His thesis does not apply, *inter alia,* to the Greco-Persian Wars (499-448 B.C., and the war between Parthia and Rome (55-38 B.C.), which were fought before the birth of Jesus (A.D. 1) or Mohammed (A.D. 570).

Ferguson (1990) offers a synergetic explanation for hostility and conflict which incorporates resource scarcity, cultural and ideological factors. His model emphasizes the interrelationship of kinships, economics and politics.[4] The persistent hostility of the Bush Administration towards the Iranian government are reflected in the following issues:

1. Efforts by the Iranian scientists to produce nuclear weapons.
2. Iranian support for international terrorist groups such as Hezbollah and Hamas.

3. Iranian policy which aims at the destruction of the state of Israel.
4. Iranian authoritarianism and human rights violations against the Iranian people.[5]

The persistent hostility of the Ahmandinejad Administration towards the American government are reflected in the following issues:

1. The exploitation of Iranian oil reserves by American oil companies in collaboration with the regime of Mohammed Reza Shah Pahlavi (1941-1979).[6]
2. Clandestine operations conducted by the Central Intelligence Agency (C.I.A.) and other American operatives against Iran, as well as economic sanctions and trade embargoes imposed against Iran.
3. American economic and military support of Israel.
4. The double standard of American leaders toward the development of Iranian nuclear energy; while, the United States maintains one of the most powerful nuclear arsenals in the world.

National pride, ethnocentrism and xenophobia on the part of some Iranians and Americans contribute to feelings of mistrust and hostility.

METHODOLOGY

The design of this study is to measure the mutual hostility between American and Iranian leaders between 2005 and 2008, analyze that hostility, and then make recommendations for detente and rapprochement between the two countries. In order to measure hostility, this research employs a multimethodical design which includes content analysis, and correlation coefficient in order to quantify, operationalize and assess the intensity of hostility between Iranian and American administrations. The study then relates the correlation coefficient with the Mirror Image formula.

Content analysis is a technique used in the study of communications-related materials and behaviors. This procedure enables the transformation of verbal and written phrases in speeches into quantitative data.[7] The paradigm selected for this task is the presidential control model of policy that views presidents as caretakers of national interests, who are Heads of State and official spokespersons for their respective administrations.[8] Speeches made by President Bush on six different occasions, and speeches made by President Ahmandinejad on three different occasions constitute the material of this study. They all refer to each other's Administration. Researchers selected 100 phrases from each of the two leaders[9] Table 2.1 gives the specific speeches utilized in this experiment.

The recorded units are themes, and the system of enumeration is the intensity of hostility which the themes represent. The reliability of the coding process is enhanced through the use of several coders (judges) seven volunteers from a graduate International Relations class during the fall 2008 semester, and seven volunteers from a graduate class on American National Defense Policy during the spring 2009 semester, judged the themes. The disagreement between the findings of the two classes was less than eight percent, and average values were selected.

MEASUREMENT

The Q-sort is a technique employed to measure the intensity of evaluative statements. Coders assigned a value of 1 to 9 to each masked phrase, according to their opinions of the level of intensity. The sources were masked to minimize bias, but later identified by suitable codes. Judges Q-sorted the 200 phrases according to a scale of hostility ranging from 1 (least hostile) to 9 (most hostile). The phrases were forced into an array, and then evaluated on the one to nine scale. Table 2.2 gives the hostility values of speeches by Presidents Bush and Ahmandinejad from 2005-2008.

It is useful to explore the mutuality of hostility between the Bush and Ahmandinejad administrations. Spearman's Rho correlation coefficient is a coefficient of association between two interval variables, measuring the closeness of fit of data points around the regression line. Like other coefficients it ranges from -1.00 (perfect negative correlation)to $+1.00$ (perfect positive correlation), with a 0.00 value indicating no association at all. Precise probabilities may be attached to its values for specified numbers of ranked pairs. The computational pattern consists of attaching a rank number to each item in each of two paired distributions, squaring the differences between each of the paired ranks, and applying a formula to the sum of these squared differences.[10] Table 2.3 gives a ranking of the most common to the least common hostile phrases of presidents Bush and Ahmandinejad. Table 2.4 gives the Mirror Equation for concave mirrors. The Mirror Equation expresses the quantitative relationship between the object distance (p), the image distance (q), and the radius of the curvature (r). Spearman's Rho merely gives a first indication of correlation, but does not relate the Mirror Equation with the mirror image theory.

However, a hostility index can be considered as a distance on an arbitrary scale. If the hostility is mirrored in a plane (flat) mirror, and if the hostility escalates proportionately, then that reality can be shown by p versus q. The regression coefficient would demonstrate the relative degree of escalation on each side of the index of regression is robust. Pearson's rho can be used to show how clearly correlated the hostility is. However, this would only be meaningful if p and q are the hostility indices of chronologically equivalent speeches.

If the points lie on a straight line then there is a direct mirroring, with a slope of 1, meaning equal degrees of escalation of hostile phrases or statements. This corresponds to the Mirror Formula with an infinite radius of curvature. If the points do not essentially lie on a straight line, then the relative escalations of hostile phrases or sentences are not proportional and a new formula must be developed. The best fit line crosses the 1/p axis gives a value of y such that $r = 2/y$ (see figure 2.1). This would then indicate a "lens" effect such that mirrored hostility is not lineally escalated Table 2.5 gives the formula for the mirror image.

ANALYSIS

Table 2.2 shows a high level of hostility existed between the Administration of Presidents Bush and Ahmandinejad. Forty-five percent of the phrases of both leaders were at the two highest intensity levels; whereas, only eight percent of their phrases were in the two lowest intensity levels. Table 2.3 demonstrates a perfect positive correlation in the ranks of hostile phrases of the two leaders. This positive correlation is compatible with the concept of mirror image theory. Mirror image is a microcosmic theory based on the assumption that the leaders of two countries involved in a prolonged hostile confrontation develop fixed, distorted attitudes about each other that are quite similar.[11]

A close examination of insinuations in the speeches of Presidents Bush and Ahmandinejad reveal an underlying conflict of national interests between Iran and the United states. In his speech on the Middle East at the University of South Carolina on May 9, 2003, Bush said: "Progress will require increased trade, . . . I propose the establishment of a U.S.-Middle East free trade area within a decade."[12] Referring to the United States and its allies during a speech before the U.N. General Assembly on September 26, 2007, Ahmandinejad asserted: "in their view, human rights are tantamount to profits for their companies. . . the plundering of the peoples' wealth by big powers. . . political and economic domination of certain powers."[13] These phrases and similar phrases indicate a significant economic motive for the conflict between Iran and the United States.

What is the cause of this economic conflict? It is important to note that of the 1,238.892 billions of barrels of known oil reserves worldwide, 755.325 billions of barrels are located in Near Eastern and Middle Eastern countries.[14] This is sixty-one percent of the known world oil reserves. Diplomatic moves by Iranian leaders imply that they seek local preponderance among members of the Organization of Petroleum Exporting Countries (OPEC). For example, On April 23, 2009, Iranian Deputy Oil Minister Noureddin Shahnazizadeh reached an agreement with Iraqi officials to construct an oil pipeline from the southern Iraqi city of Basra, to the city of Abadan in southwestern Iran.[15] In addition, between 2005 and 2008 President Ahmandinejad established close diplomatic and economic relations with President Hugo Chavez Frias of Venezuela. Venezuela is also a member of OPEC. Chavez and Ahmandinejad have signed multiple economic accords, including agreements involving oil exploration and distribution.[16]

The United States currently relies on petroleum from the Middle East and Near East to support its economy. The need for petroleum is so critical that American officials will render economic and military aid to regimes in the region which are friendly to the United States, and will militarily confront insurgencies in Iraq and Afghanistan by groups which pose a security or economic threat to American national interests. Therefore, American and Iranian national interests in the Middle East and Near East are diametrically opposed to each other. Differences in perspective over the status of Israel, and civilizational differences between Iran and the United States exacerbate the conflict which stems from control over the status of the oil reserves.

CONCLUSION

This paper is driven by the theoretical perspective of the mirror image issue, and attempts to add to the theoretical base of the mirror image through synthesizing the correlation coefficient with the Mirror Equation for concave mirrors. The concept of the mirror image in international relations was logically related to number of suggestions put forth for reducing the hostility of the Cold War, and the outright war between the superpowers, through unilateral initiative by one side, designed to reduce international tensions and evoke reciprocal gestures of cooperation from the other side. Perhaps closer ties can be forged between the leaders of Iran and the United States by the reapplication of this concept. The Obama Administration may be capable of making this adjustment in American foreign policy.

This essay has used a multidisciplinary approach to the study of mirror image. This study has sought to make a correlation between the physical and social sciences by equating the Mirror Equation with the mirror image, which is a microcosmic theory of conflict employed in international relations. A comparison of the mirror image and Mirror Equation can help scholars avoid the pitfalls of confining studies to any particular discipline.

AFTERWORD

Those scholars unfamiliar with "mirror image theory" in international relations theory, consult: Arthur Gladstone, "The Conception of the Enemy," JOURNAL OF CONFLICT RESOLUTION, III (June 1959), 132; Urie, Bronfenbrenner, "The Mirror Image in Soviet-American Relations: A Social Psychologist's Report," JOURNAL OF CONFLICT RESOLUTION, XI (September 1967), 325-332; Charles E. Osgood, "Analysis of the Cold War Mentality," JOURNAL OF SOCIAL ISSUES, XVII (3) (1961), 12-19.

Table 2.1. Sources of Speeches by Presidents George W. Bush and Mahmoud Ahmandinejad

U.S., Department of State, Assistant Secretary for Public Affairs, President George W. Bush, State of the Union Addresses: 2005, February 2, 2005; 2006, January 31, 2006; 2007, January 23, 2007; 2008, January 28, 2008, Washington, D.C.

U.N., Director, Information Centre, President Mahmoud Ahmandinejad's speeches before the United Nations General Assembly, September 20, 2006, September 26, 2007, Washington, D.C.

"Text of Mahmoud Ahmandinejad's Speech," translation by Nazila Fathi in the New York Times Tehran bureau, of October 26, 2005 to an Islamic Student Association conference on "The World Without Zionism," The conference was held in Tehran, at the Interior Ministry.

Ingela Anderson, *American Political Rhetoric: A study of selected speeches by George w. Bush* (Lulea, Sweden: Lulea University of Technology, 2005), pp. 32-46.

Table 2.2. Hostility Values of Speeches by Bush and Ahmandinejad

Intensity Category	Q-Sort Percent	Expected for 100 phrases	Observed Bush	Observed Ahmandinejad
1	5	5	3	3
2	8	8	5	5
3	12	12	6	6
4	16	16	7	7
5	18	18	9	9
6	16	16	15	14
7	12	12	10	11
8	8	8	25	24
9	5	5	20	21
TOTAL	100	100	100	100

Source: Table 2.1

Table 2.3. Spearman's Rho Ranked by the Most Common to the Least Common Level of Hostile Phrases by George W. Bush and Mahmoud Ahmndinejad

Level Of Hostility			
George W. Bush Rank	Mahmoud Ahmandinejad Rank	Difference	D^2
8	8	0	0
9	9	0	0
6	6	0	0
7	7	0	0
5	5	0	0
4	4	0	0
3	3	0	0
2	2	0	0
1	1	0	0

Source: Table 2.2   Sum $D^2 = \underline{0}$
N = 9.
Spearman's Rho = 1.00 – 6 Sum D^2
                          N(N2–1)
Rho = 1.00 – 0/720 = 1.00 – 0 = 1.00 (perfect positive correlation)
Ho: the rank of hostile phrases by Bush and Ahmandinejad are not significantly related.
Alterative hypothesis: The rank of hostile phrases is significantly related.
Significance level: .05
Statistic: Spearman's Rho.

Table 2.4. The Mirror Equation

The Mirror Formula

$$\frac{1}{p} + \frac{1}{q} = -\frac{2}{r}$$

Where p is the object distance, q is the image distance, and r is the radius of curvature.

Source: Table 2.3; Paul Peter Urone, *College Physics* (Pacific Grove, California: Brooks/Cole, 2001), pp. 636-649.

C. Mirror formulas for Image location

In place of the graphical ray-tracing methods described above, we can use formulas to calculate the image location. The derivation is typical of those found in geometrical optics, and is instructive in its combined use of algebra, geometry, and trigonometry. (If the derivation is not interest to you, you may skip to the next section, where the derived formula is used in typical calculations. Be sure, though, that you learn about the sign convention discussed below.)

1. *Derivation of the mirror formula.* The drawing we need to carry out the derivation is shown in Figure 3-18. The important quantities are the object distance *p*, the image distance shown, and the sign on r will indicate whether the mirror is concave or convex. All other quantities in Figure 3-18 are used in the derivation but will not show up in the final "mirror formula."

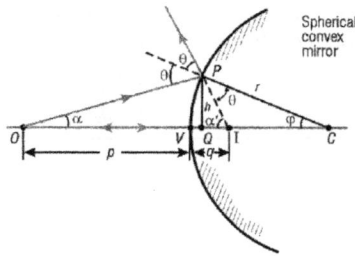

The mirror shown in Figure 3-18 is convex with center of curvature *C* on the right. Two rays of light originating at object point *O* are drawn, one normal to the convex surface at its vertex *V* and the other an arbitrary ray incident at *P*. The first ray reflects back along itself; the second reflects at *P* as if incident on a plane tangent at *P*, according to the *law of reflection*. Relative to each other, the two reflected rays diverge as they leave the mirror. The intersection of the two rays (extended backward) determines the image point *I* corresponding to object point *O*. The image is virtual and located behind the mirror surface.

Object and image distances measured from the vertex *V* are shown a *p* and *q*, respectively. A perpendicular of height *h* is drawn from *P* to the axis at *Q*. We seek a relationship between *p* and *q* that depends on only the radius of curvature *r* of the mirror. As we shall see, such a relation is possible only to a first-order approximation of the sines and cosines of angles such as α and φ made by the object and image rays at various points on the spherical surface. This means that, in place of expansions of sin φ and cos φ in series as shown here,

$$\sin \varphi = \varphi - \frac{\varphi^3}{3!} + \frac{\varphi^5}{5!} - \cdots$$

$$\cos \varphi = 1 - \frac{\varphi^2}{2!} + \frac{\varphi^4}{4!} - \cdots$$

We consider the first terms only and write sin $\varphi \cong \varphi$ and cos $\varphi \cong 1$, so that

$$\tan \varphi = \frac{\sin \varphi}{\cos \varphi} \cong \varphi$$

These relations are accurate to 1% or less if the angle φ is 10α or smaller. This approximation leads to *first-order (Gaussian) optics*, after Karl Friedrich Gauss, who in 1841 developed the foundations of his subject. Returning now to the problem at hand—that of relating *p, q,* and *r*—notice that two angular relationships may be obtained from Figure 3-18, because the exterior angle of a triangle equals the sum of its interior angles, Thus,

$$\theta = \alpha + \varphi \text{ in } \Delta OPC \quad \text{and} \quad 2\theta = \alpha + \alpha? \text{ in } \Delta OPI$$

Which combine to give

$$\alpha \,\square\, \alpha? = 2\varphi$$

Using the small-angle approximation, the angles $\alpha, \alpha?$, and φ above can be replaced by their tangents, yielding

$$\frac{h}{p} - \frac{h}{q} = -2\frac{h}{r}$$

Note that we have neglected the axial distance VQ, small when φ is small. Cancellation of h produces the desired relationship,

$$\frac{1}{p} - \frac{1}{q} = -\frac{2}{r}$$

If the spherical surface is chosen to be concave instead, the center of curvature will be to the left. For certain positions, of the object point O, it is then possible to find a real image point, also to the left of the mirror. In these cases, the resulting geometric relationship analogous to Equation 3-5 consists of the same terms, but with different algebraic signs, depending on the *sign convention* employed. We can choose a sign convention that leads to a single equation, the *mirror equation*, valid for both types of mirrors. It is Equation 3-6

$$\frac{1}{p} + \frac{1}{q} = -\frac{2}{r}$$

2. Sign Convention. The sign convention to be used in conjunction with Equation 3-6 and Figure 3-18 is as follows.

Object and image distances p and q are both *positive* when located to the *left* of the vertex and both *negative* when located to the *right*.

The radius of curvature r is positive when the center of curvature C is to the left of the vertex (concave mirror surface) and negative when C is to the right (convex mirror surface).

Vertical dimensions are positive above the optical axis and negative below.

In the application of these rules, light is assumed to be directed initially, as we mentioned earlier, from left to right. According to this sign convention, positive object and image distances correspond to real objects and images, and negative object and image distances correspond to virtual objects and images. Virtual objects occur only with a sequence of two or more reflecting or refracting elements.

3. *Magnification of a mirror image.* Figure 3-19 shows a drawing from which the magnification—ratio of image height h_i to object height h_o-can be determined. Since angles θ_i, θ_r, and α are equal, it follows the triangles VOP and $VIP?$ are similar. Thus, the sides of the two triangles are proportional and one can write

$$\frac{h_i}{h_o} = \frac{q}{p}$$

This gives at once the magnification m to be

$$m = \frac{h_i}{h_o} = \frac{q}{p}$$

When the *sign convention* is taken into account, one has, for the *general case*, a single equation, Equation 3-7, valid for both convex and concave mirrors.

$$m = -\frac{q}{p}$$

If, after calculation, the value of *m* is positive, the image is erect. If the value is negative, the image is inverted.

http://cord.org/step online/st1-3/st13eii3.htm   3/23/2009

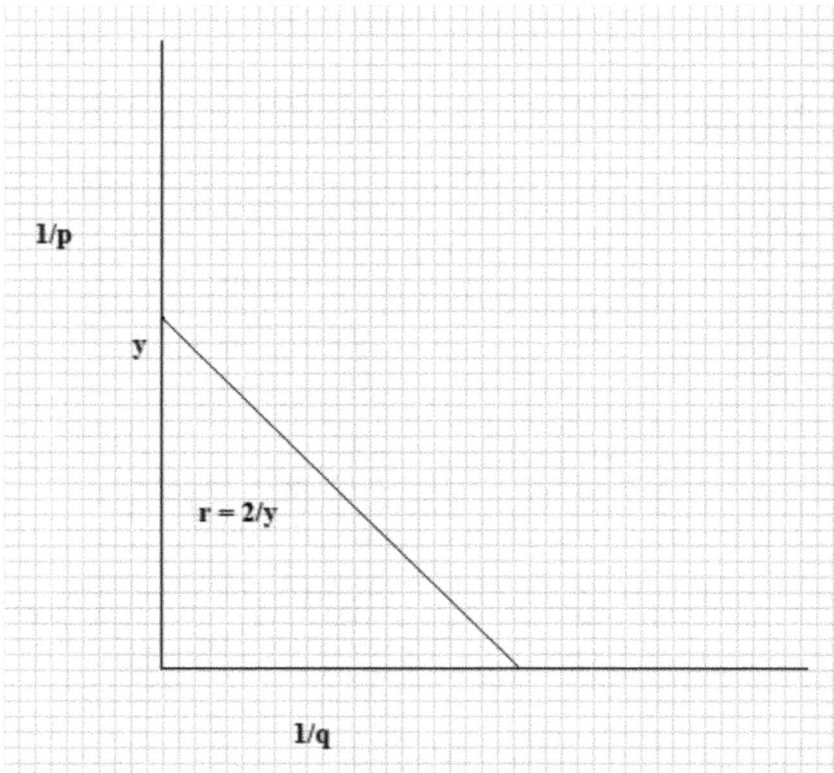

Figure 2.1.

Table 2.5.   The Mirror Image for Mutual Hostile Remarks

The Mirror Image Formula
$r = 2/y$

Where r represents the mirror image and y is the point along the axis where 1/p and 1/q meet. p represents the hostility values of phrases by President George W. Bush or Mahmoud Ahmandinejad, and q represents the hostility values of phrases in response by the other party.

Source: Table 2.3

NOTES

1. U.S. Department of State, Bureau of Near Eastern Affairs, *Background Note: Iran* (Washington, D.C., March 2008), P. 5.

2. Brian R. Ferguson (ed.), *Warfare, Culture, and Environment* (Orlando: Academic Press, 1984); p. 22, 37. For the leading economic theories see: James E. Dougherty, (et al.), *Contending Theories of International Relations* (New York: Longman, 2001), pp. 416-484.

3. Samuel P. Huntington, *The Clash of Civilizations* (New York: Simon & Schuster, 1996), pp. 308, 312.

4. Brian R. Ferguson, "Explaining War," in Jonathan Haas (ed.), *The Anthropology of War* (Cambridge: Cambridge University Press, 1990), pp. 6-43.

5. U.S., Department of State, *Background Note: Iran, op. cit.*, p. 5.

6. Lloyd Ridgeon, *Religion and Politics in Modern Iran* (New York: I.B. Tauris, 2005), pp. 175-195.

7. Janet B. Johnson (et al.), *Political Science Research Methods* (Washington, D.C.: C.Q. Press, 1986), pp. 206-210.

8. Steven W. Hook, *U.S. Foreign Policy: The Paradox of World Power* (Washington, D.C.: C.Q. Press, 2008), pp. 79-83.

9. Jaral B. Manheim (et al.), *Empirical Political Analysis* (White Plains, New York: Longman, 1990), pp. 160-161.

10. Oliver Benson, *Political Science Laboratory* (Columbus, Ohio: Charles E. Merrill, 1969), pp. 152-153.

11. Dougherty, *Contending Theories of International Relations, Op. cit.*, pp. 245-246.

12. Ingela Anderson, "President Bush Presses for Peace in the Middle East," May 9, 2003, in Ingela Anderson, *American Political Rhetoric: A Study of Selected Speeches by George W. Bush* (Lulea, Sweden: Lulea University of Technology, 2005), pp. 38-41.

13. U.N., Director, Information Centre, President Mahmoud Ahmandinejad's Speech before the United Nations General Assembly, September 26, 2007, p. 1.

14. U.S., Department of Energy, Energy Information Administration, Official Energy Statistics from the U.S. Government, Washington, D.C., March 3, 2009, p. 1.

15. George Friedman, CEO, "Iran, Iraq: An Agreement on an Oil Pipeline," *Strat for Global Intelligence*, Austin, Texas, April 24, 2009, p. 1.

16. U.S., Department of State, Bureau of Western Hemisphere Affairs, *Background Note*: *Venezuela* (Washington, D.C.: January 2009), p. 3.

Chapter Three

A Probability Experiment Involving Major American Military Conflicts (1775-2010)

"War is nothing but the continuation of policy with other means,"

—Karl von Clausewitz[1]

INTRODUCTION

Americans have mixed feelings about the U.S. military establishment. Research by Jordan (2009) shows that "to Americans, not only does normalcy refer to the primacy of domestic affairs, but it also reflects a belief that tranquility is the normal condition of the world order."[2] However, Broom (1990) demonstrates a growing confidence in the military by the American society between 1973 and 1986.[3] The social status of military personnel reached new heights after the terrorist attacks of September 11, 2001. The main purpose of this article is to examine the role of war and military operations in the conduct of American foreign between 1775 and 2010, by observation and experimentation. The key question this work addresses is whether or not war and military operations have been a primary means of accomplishing American foreign policy objectives.

The most important information in this article is the duration of 14 major armed conflicts, and the length of peace between those conflicts. The main inference in this article is that military force, or the threat of military force, has been a principal means of achieving American foreign policy goals. The key concept needed to understand this work is that the U.S. Constitution authorizes the use of military force only to protect the States ". . . against Invasion; and . . . domestic Violence."[4] This concept of the use of military action only for self-defense is reaffirmed by the United Nations Charter which allows "the inherent right of individual or collective self-defense if an armed attack occurs against a Member of the United Nations."[5] The main assumption of this article is that military force has often been used as an American foreign policy expedient beyond the necessities of self-defense.

If we take this line of reasoning seriously, the implications are a more realistic assessment of the causes of war, and through this awareness, a greater possibility of peace through diplomacy, accommodation and other means. If we fail to take this line of reasoning seriously, the implications are a continuation of military force as a primary means of accomplishing American foreign policy objectives. The main point of view presented in this work is that American foreign policy leaders can learn to more effectively employ means other than military force to obtain their goals. These means include: diplomacy, strategic intelligence, cultural exchanges, international economic policies, international law, and international conferences and organizations.

THE PATHOGENESIS OF WAR

Warfare can be viewed as armed hostilities between states or civilizations, within states or territories undertaken by means of military force.[6] There are several forms of warfare: limited or total, nuclear or nonnuclear, conventional or asymmetric, insurgencies or revolutionary and military operations against terrorist organizations. The causes of warfare are many and complex. They may include political, ideological, economic, religious and psychological factors Microcosmic theories of violent conflict focus upon factors which operate, consciously or unconsciously, at the personal level. Advocates of microcosmic theories believe the biological, psychological, social learning and personality types are partially responsible for warfare. The eminent historian John Keegan (1993) has written: "Soldiers are not as other men," they belong to "the culture of the warrior."[7]

Macrocosmic theories of warfare focus on the collective, systemic, institutional, composite or group level to discover the main issues which facilitate violent conflict.[8] Advocates of macrocosmic theories believe that disputes over national boundaries, differences in ways of living, struggles for national homelands, control of strategic places or resources, and disputes over leadership can be causes of warfare. A common macrocosmic theme is conflict over international economic issues such as trade or scarce strategic resources.[9] It is beyond the scope of this article to determine the definitive causation of warfare. However, it is possible to identify major American military conflicts between 1775 and 2010, and determine periods of war and peace.

METHODOLOGY

The design of this study is to measure in years, periods of major American wars and peace between 1775 and 2010, in an effort to determine whether or not military operations have been a primary means of accomplishing American foreign policy goals. The term "major war" means: (1) a military conflict sanctioned by the U.S. Congress by a Declaration of War;[10] or, (2) a military conflict involving more than 100,000 American combat forces. A minor conflict is one which does not meet either standard. Examples of minor conflicts include: the Whiskey Rebellion (1794), military actions in Latin America under the "Roosevelt Corollary" to the Monroe Doctrine (1904-1933), American intervention in the Dominican Republic (1965), the bombing of Libya (1986), and American military operations in Grenada (1983), Haiti (1994), Bosnia (1995), Kosovo (1999), and elsewhere.

This paper will employ the formula for empirical probability to determine the duration of major wars and periods of peace between 1775 and 2010. The study will then use a formula for subjective probability to estimate the potential for major wars and peace in the future. Finally, the article will use a formula for the number of interaction channels among different alliances to assess whether unilateral or multilateral action is best suited to prosecute war.

Probability is the chance of an event occurring. Empirical probability is a type of probability experiment that uses frequency distributions based on observations to determine numerical probabilities of events. Subjective probability is the type of probability experiment that uses a probability value based on an estimate that utilizes the findings of an empirical probability experiment.[11] The number of possible interaction channels on anyone diplomatic issue are given in the general formula for the number of combinations of things taken at a time.[12] This article contains descriptive as well as analytical information to familiarize the reader with the subject matter, and give a better contextual understanding of the empirical data.

MEASUREMENT

Table 3.1 gives the dates and duration of 14 major, American military conflicts between 1775 and 2010. Table 3.1 also shows the dates and duration of periods of peace between these 14 military conflicts. The Table demonstrates that the total number of years of warfare exceeds the total number of years of peace by 28 years. The most extended period of warfare occurred between 1817 and 1898. The most extended period of peace occurred between 1917 and 1941. The greatest number of minor conflicts took place between 1904 and 1933, as part of the Roosevelt Corollary to the Monroe Doctrine; and after World War II, as part of the Cold War and War on Terrorism.

Table 3.2 gives the formula for empirical probability. The Table demonstrates that the United States has been engaged in major wars 56.5 percent of the time between 1775-2010. Table 3.3 shows the average number of years of peace for the 14 conflicts under study is 6.6 years. Table 3.4 displays the possible number of interaction channels on any one issue involving the potential for military conflict. In the case of bilateral diplomacy there is only one interaction channel, between the United States and its opponent on any given issue. Diplomatic exchanges between two states are called bilateral diplomacy. Diplomatic exchanges among more than two states are called multilateral diplomacy.

Table 3.4 also gives the possible interaction channels involving the 15-member U.N. Security Council. In the case of the Security Council there are 105 possible interaction channels on any one issue. Table 3.4 leads to some interesting conclusions. First, the data provides a way of determining the possible number of interaction channels in any alliance or international organization on any particular issue. The greater the number of actors, the greater the number of potential interaction channels. Second, the number of possible interaction channels in the U.N. Security Council is 105 times greater than a unilateral action by the United States. Third, the potential instantaneous rate of change of potential interaction channels in the Security Council, on any particular issue, is 9.7 times greater than a unilateral action by the United States.

ANALYSIS

Contrary to the belief of many Americans, warfare rather than peace is the more common position of American foreign policy. Evidence presented in this paper demonstrates that military operations have been a principal means of accomplishing American foreign policy objectives between 1775 and 2010. Some wars appear to meet the criteria of the U.S. Constitution as necessary conflicts. Examples of wars of this kind are the War of 1812, the American Civil War (1861-1865), World War II (1941-1946), and perhaps the War in Afghanistan (2001-).

Other wars do not meet the U.S. Constitutional requirements for just military conflicts: examples of wars of this kind are the Mexican-American

War (1846-1848) or Spanish-American War (1898). These conflicts were wars of expansion in which the United States acquired vast lands in North America and abroad.[13] The Korean War (1950-1953) and Vietnam War (1964-1973) also do not meet the standards of the U.S. Constitution. These military operations were in support of the policy of containment. Containment was a general policy adopted in 1947, under the Administration of Harry S. Truman, to build positions of strength around the globe in order to contain Communist expansionism within its existing boundaries.[14] Perhaps multilateral diplomacy would have produced better results than military operations.

Tables 3.2 and 3.3 demonstrate that war was more likely than not between 1775-2010 and that there were less than seven years peace between major military conflicts on average. Should U.S. policy-makers rely more on multilateral action through international organizations such as the United Nations or the North Atlantic Treaty Organization (NATO) in order to solve difficult international problems? Under international law the U.N. Security Council is responsible for determining "the existence of any threat to the peace, breach of the peace, or act of aggression."[15] The American government agreed to the terms of the Charter on October 24, 1945. In the event of military aggression, the Security Council "may take such action by air, sea, or land forces as may be necessary to maintain or restore international peace and security."[16]

The U.N. Charter stipulates that "decisions of the Security Council...shall be made by an affirmative vote of nine members including the concurring votes of the permanent members."[17] The five permanent members include: China, France, Great Britain, Russia, and the United States. These five countries have an absolute veto over Security Council resolutions, and may exercise their veto to block resolutions inimical to their national interests. Since few, important international issues do not adversely affect one or more of the permanent members of the Security Council it is difficult for the Security Council to fulfill its mission. The difficulty experienced by the U.S. government in applying greater, multilateral economic sanctions against Iran, because of Iran's nuclear program, is a case in point.

Table 3.4 shows the number of possible interaction channels in any one diplomatic issue to be 105 for the Security Council. The greater the number of interaction channels, the greater the potential for disagreement among the members. The instantaneous rate of change of potential interaction channels on any particular issue is 14.5. Therefore, there is a significant potential for disagreement or changes in agreement, among members of the Security Council.

Can the North Atlantic Treaty Organization (NATO) offer a better multilateral approach to international peace and security than the U.N. Security Council? The number of possible interaction channels in anyone diplomatic issue for the 28 country alliance is 378. The instantaneous rate of change of potential interaction channels on any particular issue is 27.5. Therefore, the potential for disagreement or changes in agreement, is greater among the countries of the NATO alliance than the countries of the Security Council. In addition, there are more common interests inherent in the NATO alliance compared to the more disparate membership of the Security Council.

For an alliance to be operative, individual, national leaders must perceive that their national interests are best served by complying with the terms of the alliance.[18] Some nations may honor their collective obligations, others might stand aside, and some might support an aggressor's actions.[19] The Iraq War (2003-) provides an example. Members of NATO responded differently to the conflict. For example, the United Kingdom, Poland and Romania supported the U.S. Conflict,[20] France and Germany did not join the coalition.[21] Spain withdrew its forces on April 17, 2004.[22] Turkey refused to allow American troops, and logistical supplies, to pass through its territory in March 2003.[23]

CONCLUSION

American leaders can learn to more effectively employ means other than military force to achieve their foreign policy objectives. Diplomacy, intelligence estimates, cultural exchanges, international economic policies, international law, and international conferences and organizations often can be viable alternatives to the use of force. Examples are the American-North Vietnamese peace talks, held in Paris and the United States-Soviet Strategic Arms Limitation Talks (SALT) held at Vienna, Helsinki, and Geneva. On April 8, 2010, President Barack Obama and Russian President Dmitriy Medvedev signed a new Strategic Arms Reduction Treaty in Prague, with both leaders pledging to reduce their deployed, strategic nuclear weapons stockpiles. Mutual respect can be achieved through international exchanges of scholars and students, such as the Fulbright Scholarship program, or through sports, such as the International Olympics, or International Military Sports Council. However, the findings of this study demonstrate that warfare and military operations have been a primary means of accomplishing American foreign policy objectives between 1775 and 2010.

A liberal theory of international relations asserts that many countries have adopted political democracy and a market economy. Therefore, these factors are becoming increasingly important and military factors less important. This point of view may be correct in the future. However, wars and the threat of military conflicts are current realities that must be addressed.

Microcosmic and macrocosmic factors too complex to thoroughly understand, make it more likely than not, that military operations will continue to be an important instrument of American foreign policy in the future. To be effective, American alliances should be comprised of a small number of nations whose leaders perceive that their national interests coincide with an American victory. An example is the American Revolution (1775-1783) which was prosecuted with French, Dutch and Spanish financial and military assistance. In contrast, the 28 members of the NATO alliance contained too many interaction channels, and potential disagreements, for that organization to be an effective means for prosecuting Operation Iraqi Freedom.

A smaller number of allies confer advantages in terms of freedom of action. On the other hand, a large alliance can involve greater economic and military resources than a smaller alliance. The U.S. Presidency and Pentagon must weigh the advantages and disadvantages of the size of an alliance to meet the particular military threat posed in different theaters of operation, as part of the regional defense strategy.

Table 3.1.   Major American Military Conflict (1775-2010)

Wars	Dates	Duration	Periods Of Peace	Duration
American Revolution	1775-1783	8 yrs	1783-1801	18 yrs
Tripoli War	1801-1805	4 yrs	1805-1812	7 yrs
War of 1812				
Tripoli War	1812-1815	3 yrs	1815-1817	2 yrs
Indian Wars				
Mexican				
American War	1817-1898	81 yrs	1898-1917	19 yrs
American Civil War				
Spanish-American War				
WW I	1917-1918	2 yrs	1918-1941	23yrs
WW II	1941-1946	5 yrs	1946-1950	4 yrs
Korean War	1950-1953	3 yrs	1953-1964	11 yrs
Vietnam War	1964-1973	9 yrs	1973-1990	17 yrs
Persian Gulf War	1990-1991	1 yr	1991-2001	10 yrs
Iraq and Afghanistan	2001-2010	9 yrs		
TOTAL		121 yrs		93 yrs

Source: U.S., Department of Defense, Assistant Secretary for Public Affairs, Washington D.C., January 21, 2010.

Table 3.2. Formula for Empirical Probability

Formula

$$P = \frac{f}{n}$$

where: P = the probability of war
f = the duration of wars in years (1775-2010)
n = the total number of years (1775-2010)

$$P = \frac{121}{214}$$

$$P = 0.565 \text{ or } 56.5\%$$

Sources: Table 3.1; Allan G. Bluman, *Elementary Statistics: A Step by Step Approach* (Boston: McGraw-Hill, 2007), pp. 181-183.

Table 3.3. Formula for Subjective Probability

Formula

$$D = \frac{p}{n}$$

where: D = the average number of years of peace between wars
p = the total number of years of peace
n = the number of wars under study

$$D = \frac{93}{14}$$

$$D = 6.6 \text{ years}$$

Sources: Table 3.1; Allan G. Bluman, *Elementary Statistics: A Step by Step Approach* (Boston: McGraw-Hill, 2007), pp. 181-183.

NOTES

1. Karl von Clausewitz, *On War*, translated by O.J. Mathhias Jolles (New York: Random House, 1943), p. 69

2. Amon A. Jordan (et al.), *American National Security* (Baltimore, Maryland: The Johns Hopkins University Press, 2009), p. 27.

3. Leonard Broom (et al.), *Sociology* (Belmont, California: Wadsworth, 1990), p. 165.

4. U.S., *Constitution*, Art. IV, sec. 4.

5. U.N., *Charter*, Chapter 7, art. 51.

6. Jack C. Plano (et al.), *Political Science Dictionary* (Hinsdate, Illinois: The Dryden Press, 1973), p. 409.

7. James E. Dougherty, *Contending Theories of International Relations* (New York: Longman, 2001), pp.231-254; John Keegan, *A History of Warfare* (New York: Vintage Books, 1993), p. xvi.

Table 3.4. Number of Possible Interaction Channels on Any One Diplomatic Issue

Formula

$$y = \frac{x^2 - x}{2}$$

where: x = the number of national actors
y = the number of possible interaction channels

Bilateral Diplomacy:

$$Y = \frac{2^2 - 2}{2}$$

$$Y = 1$$

$$\frac{dy}{dx} = x - \tfrac{1}{2}$$

$$\frac{dy}{dx} = 1.5$$

Security Council Involvement:

$$Y = \frac{15^2 - 15}{2}$$

$$Y = 105$$

$$\frac{dy}{dx} = 15 - \tfrac{1}{2}$$

$$\frac{dy}{dx} = 14.5$$

Sources: The general formula for the number of combinations of n things taken r at a time; basic equation of differential calculus; in William M. Setek, *Fundamentals of Mathematics* (New York: Macmillan, 1989), pp. 167-169.

8. Dougherty, *Contending Theories of International Relations*, pp. 264-321.

9. *Ibid.,* pp. 416-483.

10. U.S. *Constitution*, Art. I, sec. 8.

11. Allan G. Bluman, *Elementary Statistics: A Step by Step Approach* (Boston: McGraw-Hill, 2007), pp. 181-185.

12. William M. Setek, *Fundamentals of Mathematics* (New York: Macmillan, 1989), pp. 167-169.

13. Joseph R. Conlin, *The American Past: A Survey of American History* (New York: Harcourt College Publishers, 2001), pp. 393-397; 649-654.

14. Jordan, *American National Security, Op. cit.*, pp. 44-45.

15. U.N., *Charter*, Chapter 7, art. 39.

16. U.N., *Charter*, Chapter 7, art. 42.

17. U.N., *Charter*, Chapter 7, art. 17, par. 3.

18. Hans J. Morganthau, *Politics Among Nations* (New York: Alfred A. Knopf, 1978), p. 193.

19. *Ibid.,* p. 421.

20. U.S., Department of State, Bureau of European and Eurasian Affairs, *Background Note: United Kingdom*, No. 3846, Washington, D.C., March 2009, p. 7.

21. U.S., Department of State, Bureau of European and Eurasian Affairs, *Background Note: France*, No. 3842, Washington, D.C., April 2009, p. 9.

22. U.S., Department of State, Bureau of European and Eurasian Affairs, *Background Note: Spain*, No. 2878, Washington, D.C., June 2009, p. 4.

23. U.S., Department of State, Bureau of European and Eurasian Affairs, *Background Note: Turkey*, No. 3432, Washington, D.C., May 2009, p. 7.

Chapter Four

Radioactive Decay and the Rate of Decline of Empires: A Connection between the Physical and Social Sciences

"Let Rome in Tiber melt, and the wide arch of the rang'd empire fall!"

—William Shakespeare, *Antony and Cleopatra*, Act I, scene 1, 33.

INTRODUCTION

The main purpose of this article is to demonstrate a correlation between the decay of radioactive substances and the rate of decline of empires. The key question the author is addressing is-does the radioactive decay curve employed in physics correspond to the rate of decline of empires? The most important information in this article is the constant for the radioactive decay curve, and the rate of decline of seven empires under study. The main inference in this article is that an interdisciplinary approach to the decline of empires can challenge the common narrowness of ingrained, discipline-bound ways of thinking that produces occlusion as well as insight.

The key concept needed to understand this article is-If the knowable universe is a unity, discipline is a loss as well as a gain, and interdisciplinary, trans-discipline is or multidisciplinary foci can effectively cross disciplinary boundaries to enhance understanding of the rate of decline of empires. The main assumption underlying the author's thinking is that common approaches used to study physics and political science may need to be balanced with and measured against each other in a broader perspective in order to better understand the rate of decline of empires. If we take this line of reasoning seriously, the implication is a reconstructive, holistic connection between and among the physical and social sciences, and an escape from the occlusions of narrow methodological approaches to knowledge. If we fail to take this line of reasoning seriously, the implication is at times, to leave distinctive disciplinary modes stranded, and separated from their natural roots and social ends.

THE EMERGENCE AND EXPANSION OF EMPIRES

An empire may be defined as "the dominion or jurisdiction of an emperor; the region over which the dominion of an emperor extends; imperial power; supreme dominion; sovereign command."[1] This study is limited to those sovereign dominions led by a supreme ruler, titled emperor. For an empire to continue beyond the life of its founder, usually a person of great, charismatic personality, the empire must be based upon a civilization, united by a common ideal. Huntington (1997) defines civilization as "the highest cultural grouping of people and the broadest level of cultural identity people have short of that which distinguishes humans from other species. It is defined both by common objective elements, such as language, history, religion, customs, institutions, and by the subjective, self-identification of people."[2]

The empires forged by Alexander the Great (336-323 B.C.), and Augustus Caesar (27 B.C.-A.D. 476) provide contrasting examples about the longevity of empires. Alexander's empire did not outlast his life. After his death in 323 B.C. it split into smaller units, led by the descendants of (Ptolemy, Seleucus and Antigonus.[3] Augustus' empire lasted for centuries after the death of Augustus in A.D. 14. The reason is the Roman conquerors transformed the conquered into Romans either by accepting them into the dominant civilization as citizens or client states, or by uprooting them from their native civilizations and making them into slaves. The Romans also transformed themselves by remaking their own civilization in the image of the civilizations of the conquered. This was especially true of the Eastern Roman Empire (Byzantine Empire) (330-1453).[4] Through the process of mutual adaptation, acculturation and assimilation, the Romans created a new moral and political community coextensive with its conquests and capable of lending stability to the empire.[5] In addition, Rome was superior in political acumen and military strength to any one of the component parts of the Empire. Legionnaires could suppress revolts, insurgencies and uprisings through military force.[6]

THE DECLINE AND FALL OF EMPIRES

Several studies demonstrate the obvious-empires collapse when the civilizations of which they are a part decline and fall.[7] According to Quigley

(1961) civilizations transform through seven stages: mixtures, gestation, and expansion, age of conflict, universal empire, decay and invasion.[8] There are macrocosmic and microcosmic theories which can be applied to the decay (decline) of empires. Macrocosmic theories examine groups, collectivities, social institutions, social classes, large political movements, religious or ethnic entities, nation-states, coalitions, and cultural or global systems. Microcosmic theories focus on the behavior of individuals.[9] This author believes that both macrocosmic and microcosmic factors are involved in the decline and fall of empires.

There are three basic causes for the fall of empires. The first cause is the removal of an emperor by coup d'état or revolutionary force opposed to the empire. An example is the republican revolution led by Sun Yat-sen which forced the last Chinese Emperor, Henry Pu-yi, to abdicate in 1912.[10] A second cause for the fall of an empire is through decay, followed by a successful invasion from outside the empire. The Western Roman Empire ended when Odoacer, a powerful German chieftain, deposed Emperor Romulus Augustulus in A.D. 476.[11] A third way an empire can come to an end is by the decay and disintegration of the civilization which supports it. Sometime in the eighth or ninth century, classical Mayan civilization in the central Yucatan peninsula began to decay. It eventually ceased to exist. Most scholars attribute the collapse of Mayan civilization to over cultivation of the land by a growing population, and a drought which lasted for almost two hundred years.[12]

The author offers the following macrocosmic considerations for the decay (decline) of the empires included in this study

1. A decline in the ability of urban areas to serve as centers for political, economic, social, cultural, and religious development.
2. A diminishment of a distinct religious structure.
3. A lessening of the political and military structures to perform their responsibilities in an effective way.
4. A dwindling of social structures and economic viability.
5. A general decline in the writing ability of the general population.
6. A significant decline in artistic and intellectual activity.[13]

This author offers the following microcosmic considerations for the decay (decline) of the empires included in this study:

1. Fear of clear and present dangers, and anxiety caused by anticipated invasions, epidemics, famine, or other threats.
2. Lack of confidence in, and support for the political system and its leaders.

3. Vices and corruption displacing core virtues and values.
4. A marked decline in morale and social capital.
5. Increased cynicism, disillusionment, and the fear-induced aggression.
6. Boredom and apathy.[14]

These macrocosmic and microcosmic considerations can be caused or exacerbated by several factors. Some of these factors include: political corruption, totalitarianism, military threats and invasions, civil wars, ethnic, religious or class conflicts, pandemic diseases, natural disaster, prolonged droughts or floods, widespread contamination of food supplies or sources of water, uncontrolled immigration of peoples with alternative, primitive, core cultures, and economic extortion, sanctions or indebtedness to nations or states outside of the empire.[15]

METHODOLOGY

This study is designed to measure the rate of decline of seven Western empires. Investigation was limited to empires associated with Western civilization to avoid variables associated with civilization differences suggested by Samuel Huntington (1997). Huntington identifies eight major civilizations: Western, Confucian, Japanese, Islamic, Hindu, Slavic-Orthodox, Latin American, and possibly African.[16] Huntington includes Europe, North America and other countries settled by Europeans, such as Australia and New Zealand, as parts of Western civilization.[17] Hanson (2009) defined the concept of the West as "the culture that originated in Greece, spread to Rome, permeated Northern Europe, was incorporated by the Anglo-Saxon tradition, spread through British expansionism, and is associated today primarily with Europe, the United States, and the former commonwealth countries of Britain."[18]

This paper employs a multi-methodological scheme which utilizes an adaptation of the radioactive decay formula for radioactive substances, used in physics, to measure the rate of decline of seven Western empires. The study then applies a correlation coefficient to compare the half-life of the empires with their rates of decline, in order to determine if these two factors are significantly related. Table 4.1 gives the radioactive decay formula for radioactive substances. In the formula, (A) represents the decay constant. The decay constant is equal to the natural logarithm of two (approximately 0.69315) divided by half-life in seconds. The half-life ($t\frac{1}{2}$) can vary tremendously depending on the particular radioactive substance involved.[19] Table 4.2 gives the rate of decline formula for extinct empires. The formula is an adaptation of the radioactive decay formula. In this formula, (D) represents he decline constant. The decline constant is equal to the natural logarithm of two (approximately 0.69315) divided by the half-life in years. The half-life ($t\frac{1}{2}$) can vary significantly depending on the particular empire involved. The formula is only useful for measuring the decline constant of extinct empires, since it is impossible to determine the longevity or half-life of ongoing empires.

MEASUREMENT

Table 4.3 gives the rates of decline of seven empires whose capitals were located in Europe. The empires are chronologically listed according to the dates of their emergence. The dates are interpretative and meant to be suggestive. The table lists the dates the empires were in existence, their half-lives, and rates of decline according to the formula developed in Table 4.2. Table 4.4 offers the half-lives and rates of decline of the seven empires under study. Seven represents the most long-lived empire, and one the most short-lived empire; and 7 represents the fastest rate of decline and 1 the slowest rate of decline.

Table 4.4 also employs Spearman's Rho correlation coefficient. Spearman's Rho is a coefficient of association between two interval variables, measuring the closeness of fit of data points around a regression line. In this study it is designed to measure the half-life of empires with their rates of decline. Like other coefficients it ranges from -1.00 (perfect negative correlation) to +1.00 (perfect positive correlation), while a 0.00 value indicates no association at all. Precise probabilities may be attached to its values for specified numbers of ranked pairs. The computational pattern consists of attaching a rank number to each item in each of two paired distributions, squaring the differences between each of the paired ranks, and applying a formula to the sum of these squared differences.[20]

ANALYSIS

Spearman's Rho demonstrates a perfect negative correlation between the half-lives of the empires and their rates of decline. The null hypothesis, which states that the half-lives and rates of decline of the empires are not significantly related, is rejected. The alternative hypothesis, which states that the two factors are significantly related, is accepted. The data shows that the greater the half-life of an empire, the slower its rate of decline. The Byzantine Empire (Eastern Roman Empire) had the longest half-life, and slowest rate of decline. The Austro-Hungarian Empire had the shortest half-life and fastest rate of decline.

The Byzantine Empire (330-1453) owed its longevity to two important factors. The first factor was the mutual adaptation, acculturation and assimilation of its core culture with the core cultures of the lands it occupied. A strong cultural unity evolved from this syncretic process that was Greek and Eastern Orthodox, with Asian features and characteristics. However, the Byzantines preserved many of the customs and traditions of the Western Roman Empire, and continued to consider themselves as Romans.[21]

People from other cultures and civilizations who settled in Byzantium were referred to as Romans. For example, the great Persian poet, Jalal'uddin Mowlavi (1207-1273), became known as Rumi (the Roman), after he settled in Iconium (Konya, Anatolia) within the empire between 1215-1220.[22] The ideal of ancient Rome also crystallized in Western Europe in the form of the Holy Roman Empire (962-1806). The second factor which accounts for the longevity of the Byzantine Empire was its effective government, economic vitality, and military preparedness. The strategic location of its capital at Constantinople, protected by the sea and walled fortifications, allowed the city to withstand numerous Germanic, Persian and Arabic attacks, until its collapse in 1453.[23]

The Austro-Hungarian Empire (1867-1918) collapsed after a half-life of only 25.5 years. Initially, the charismatic personality of Emperor-King Francis Joseph, its military forces, elaborate bureaucracy and security considerations, held the empire together. However, the empire had neither a cultural syncretism around a common ideal, nor sufficient military strength to be a long-lived empire. The impact of World War I (1914-1918), and the nationalistic aspirations of Czechs, Poles, Serbs, Italians and Rumanians, led to the dissolution of the empire in 1918.[24]

CONCLUSION

This study has used a multi-methodological design, which adapts the radioactive decay formula employed in physics, to measure the rate of decline of seven extinct empires. Spearman's Rho demonstrates a perfect negative correlation between the half-life of an empire and its rate of decline. The greater the half-life of an empire, the slower will be its rate of decline. Extensive research implies that cultural syncretism based on a common ideal, and military preparedness are the main factors which influence the longevity of empires. The methodology employed in this project can be applied to the study of extinct empires worldwide.

Table 4.1.   Radioactive Decay Formula for Radioactive Substances

Radioactive Formula
$A = 0.69315/t^{1/2}$

Where: A is the decay constant. The decay constant is equal to the natural logarithm of 2 (approximately 0.69315) divided by the half-life seconds.

Source: Jearl Walker, *Fundamentals of Physics* (Hoboken, N.J.: John Wiley & Sons, 2008), pp. 1174-1177, 1219-1220.

Table 4.2.   Rate of Decline Formula for Extinct Empires

Rate of Decline Formula
$D = 0.69315/t^{1/2}$

Where: D is the rate of decline constant. The rate of decline constant is equal to the natural logarithm of 2 (approximately 0.69315) divided by the half-life of the empire in years.

Source: The formula is an adaptation of the radioactive decay formula given in Table 4.1.

Table 4.3.   Rate of Decline of Extinct Empires

Empire	*Date in Existence**	*Half-life*	*Rate of Decline*
Etruscan	753-396 B.C.	178.5	.0039
Roman (West)	27 B.C.-A.D 476	251.5	.0028
Macedonian	338-200 B.C.	69.0	.0100
Byzantine	330-1453	561.5	.0012
Holy Roman	962-1806	422.0	.0016
Hapsburg	1282-1918	318.0	.0022
Austro-Hungarian	1867-1918	25.5	.0272

* Dates are interpretative and meant to be suggestive.
Source: Felip Fernandez-Armesto, *The World: A History* (Upper Saddle River, N.J.: Pearson, 2007), pp. 135-136, 203-204, 333-334; William J. Duiker, *World History* (Belmont, CA: Wadsworth, 2004), pp. 121-123, 109-117, 131-147; C. Harold King, *A History of Civilization* (New York: Houghton Mifflin, 2004), pp. 590-591, 938-939; and other more specialized books and articles.

NOTES

1. Henry Campbell Black, *Black's Law Dictionary* (St. Paul, Minn.: West Publishing Company, 1979), p. 471.

2. Samuel P. Huntington, *The Clash of Civilization* (New York: Touchstone, 1997), p. 43.

3. D. Brendan Nagle, *Readings in Greek History* (New York: Oxford University Press, 2007), pp. 246-266.

Table 4.4.   The Half-Life and Rate of Decline of Empires*

Empires	Half-Life	Rate of Decline	Difference	D^2
Byzantine	7	1	6	36
Holy Roman	6	2	4	16
Hapsburg	5	3	2	4
Roman (West)	4	4	0	0
Etruscan	3	5	−2	4
Macedonian	2	6	−4	16
Austro-Hungarian	1	7	−6	36

Source: Table 4.3   Sum D2 =   112

N = 7.   Spearman's Rho $=$ $1.00 - \dfrac{6 \text{ Sum D2}}{N(N2\text{-}1)}$

$=$ $1.00 - \dfrac{672}{336}$

$=$ $1.00 - 2.00 = -1.00$ (perfect negative correlation)

* Where 7 represents the most long-lived empire, and 1 the most short-lived empire; and 7 represents the
  fastest rate of decline, and 1 the slowest rate of decline.
Alternate hypothesis: The two factors are significantly related.
Significance level: .05
Statistic: Spearman's Rho.
On the basis of the computation, the null hypothesis is rejected, and the alternate hypothesis is accepted

    4. Mary T. Boatwright (et al.), *The Romans: From Village to Empire* (New York: Oxford University Press, 2004), pp. 267-316, 454-457.

    5. Hans J. Morganthau, *Politics Among Nations* (New York: Alfred A. Knopf, 1978), p. 504.

    6. *Ibid.*, p. 87.

    7. For example-Carroll Quigley, *The Evolution of Civilizations: An Introduction to Historical Analysis* (New York: Macmillan, 1961; Matthew Melko, *The Nature of Civilizations* (Boston: Porter Sargent, 1969); Arnold J. Toynbee, *A Study of History* vol. IX (London: Oxford University Press, 1954), pp. 250-255.

    8. Quigley, *Ibid.*, pp. 127, 164-166.

    9. James E. Dougherty (et al.), *Contending Theories of International Relations* (New York: Longman, 2001), p. 192.

    10. C. Harold King, *A History of Civilization* (New York: Charles Scribner's Sons, 1969), p. 812.

    11. *Ibid.*, pp. 162.

    12. William J. Duiker (et al.), *World History* (Belmont, CA: Thompson/Wadsworth, 2004), pp. 162-163.

    13. *Ibid.*, pp. 7-8.

    14. Kenneth Clark, *Civilization: A Personal View* (New York: Harper & Row, 1969), pp. 1-31; Francis Neilson, "The Decline of Civilizations," *American journal of Economics and Sociology*, Vol 4, no. 4, July 1945, pp. 479-497; John M. Darley (et al.), *Psychology* (Englewood Cliff, N.J.: Prentice-Hall, 1984), pp. 345-346; Jack C. Plano (et al.), *Political Science Dictionary* (Hinsdale, Illinois: Dryden Press, 1973), p. 222.

15. There are numerous specialized books and articles which address the impact of these factors upon the people, institutions and viability of empires, and the civilizations upon which they depend. For scholars and students merely interested in an overview of these factors, consult-John P. McKay (et al.), *A History of World Societies* (New York: Houghton Mifflin, 2004); Felip Fernandez-Armesto, *The World: A History* (Upper Saddle River, N.J.: Pearson/Prentice Hall, 2007).

16. Huntington, *The Clash of Civilizations, op. cit.*, pp. 40-48.

17. *Ibid.*, p. 46.

18. Victor Davis Hanson, "The Future of Western War," *Imprimis*, Vol. 38, no. 11, November 2009, pp. 1-2.

19. Jearl Walker, *Fundamentals of Physics* (Hoboken, N.J.: John Wiley & Sons, 2008), pp. 1174-1177.

20. Oliver Benson, *Political Science Laboratory* (Columbus, Ohio: Charles E. Merrill, 1969), pp. 152-153.

21. W. Goffart, *Barbarians and Romans: The Techniques of Accommodation* (Princeton, N.J.: Princeton University Press, 1980), pp. 211-230.

22. John Moyne and Coleman Barks, *Open Secret: Versions of Rumi* (Putney, Vermont: Threshold Books, 1984), p. xi.

23. M. Whittow, *The Making of Byzantium, 600-1025* (Berkeley: University of California Press, 1996), pp. 99-103.

24. Arthur Rosenberg, *Imperial Germany, The Birth of the German Republic, 1871-1918* (Boston, MA.: Beacon Press, 1967), pp. 20-129.

Conclusion

"Life is the art of drawing sufficient conclusions from insufficient premises."

—Samuel Butler, *Note Books*. Life, IX.

This monograph has presented four case studies which attempt to make correlations between the physical and social sciences. The traditional, empirical and postmodernist approaches to the study of the social sciences have left many scholars dissatisfied with the results of these methods. I believe that the empiricists were on the right track, but that they did not go far enough. It is important to anchor statistical data to mathematical formulae or the laws of physics, in order to minimize the conscious or unconscious bias of some scholars, who might otherwise manipulate data in support of preconceived notions. Some scholars may accept this pilot study and see to advance correlations between the physical and social sciences. Other scholars might reject this approach without reservation. I conclude with the words of Henry Wadsworth Longfellow: "Sometimes we may learn more from a man's errors, than from his virtues."

In summary, mathematical formulae and the laws of physics can take scholars further in deriving conclusions from sets of assumptions than can inferential statistics. The use of inferential statistics in the social sciences is sufficiently regular that the correlations from many mathematical formulae and physical laws prove valid. Modeling of social and political phenomena is more complex than problems in mathematics and the physical sciences, because Homo sapiens are more complicated and unpredictable than the behavior of atoms or the use of numbers and symbols.

This complexity has two implication for relationships between the physical and social science. First, correlations between the physical and social sciences should begin with the simpler and more regularly observed behaviors, and then advance to more complex behaviors. Second, the mathematics necessary to address social and physical phenomena will be more complex than the mathematics which deals with the problems of classical physics.[1] This monograph is a challenge to scholars to seek correlations between the physical and social sciences.

NOTE

1. Jarol B. Manheim (et al.) *Empirical Political Analysis* (White Plains, N.Y.: Longman, 1991), pp. 309-310.

Appendix

GASES: GRAHAM'S LAWS OF DIFFUSION AND EFFUSION

Only a few physical properties of gases depends on the identity of the gas.

Diffusion - The rate at which two gases mix.

Effusion - The rate at which a gas escapes through a pinhole into a vacuum.

Thomas Graham

GRAHAM'S LAW OF DIFFUSION

The rate at which gases diffuse is inversely proportional to the square root of their densities.

$$\text{Rate}_{\text{diffusion}} \propto \frac{1}{\sqrt{\text{density}}}$$

Since volumes of different gases contain the same number of particles (see *Avogadro's Hypothesis*), the number of moles per liter at a given T and P is constant. Therefore, the density of a gas is directly proportional to its molar mass (MM).

$$\text{Rate}_{\text{diffusion}} \propto \frac{1}{\sqrt{\text{MM}}}$$

GRAHAM'S LAW OF EFFUSION

The rate of effusion of a gas is inversely proportional to the square root of either the density or the molar mass of the gas.

$$\text{Rate}_{\text{effusion}} \propto \frac{1}{\sqrt{\text{density}}} \propto \frac{1}{\sqrt{\text{MM}}}$$

http://www.chem.tamu.edu/class/majors/tutorialnotefiles/graham.htm
6/9/2008

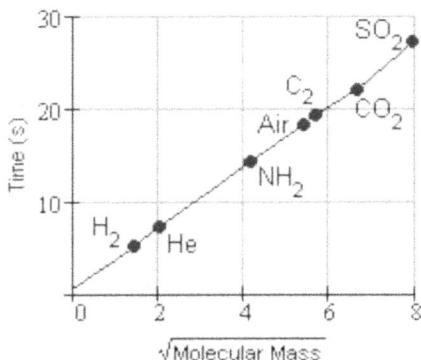

THE KINETIC MOLECULAR THEORY AND GRAHAM'S LAW

Since KE_{avg} is dependent only upon T, two different gases at the same temperature must have the same KE_{avg}:

$$\frac{1}{2} m_H v_H{}^2 = \frac{1}{2} m_O v_O{}^2$$

Simplify the equation by multiplying both sides by two:

$$m_H v_H{}^2 = m_O v_O{}^2$$

Rearrange to give the following:

$$\frac{v_H{}^2}{v_O{}^2} = \frac{m_O}{m_H}$$

Take the square root of both sides to obtain the following relationship between the ratio of the velocities of the gases and the square root of the ratio of their molar masses:

$$\frac{v_H}{v_O} = \frac{\sqrt{m_O}}{\sqrt{m_H}}$$

This equation states that the velocity (rate) at which gas molecules move is inversely proportional to the square root of their molar masses.

Next: *"Deviation from Ideal Gas Behavior: Van der Waals Equation"*
http://www.chem.tamu.edu/class/majors/tutorialnotefiles/graham.htm
6/9/2008

Index

About the Author

Dr. **Valentine J. Belfiglio** is a Professor of Government at Texas Woman's University. He also taught at the University of Oklahoma. His specialties are International Relations and American National Defense Policy.

Dr. Belfiglio has written six books and more than 100 articles. His works have been published in English and Italian. Some of them have been translated into Spanish, Hebrew, and Chinese. Dr. Belfiglio received a postdoctoral fellowship from the Foreign Affairs Association, three grants from the Texas Committee for the Humanities, and a sabbatical leave from Texas Woman's University in 2001 for research in Italy. In 1982, the President of Italy conferred upon him the title of Cavaliere dell'Ordine al Merito della Repubblica Italiana (Knighthood), for his extensive work in promoting Italian culture in the United States. In 1990, the East Texas Historical Association bestowed upon Dr. Belfiglio the C.K. Chamberlain Award for excellence in historical research and writing. In 1985, he won the Guido Dorso Prize in literature in the category of research, from the University of Naples. Dr. Belfiglio was the recipient of the Cornaro Award from Texas Woman's University for excellence in teaching and outstanding commitment to scholarship and the advancement of learning in 2003. He has presented referred papers at several international conferences.

Dr. Belfiglio was born in Troy, New York. He graduated from the Albany College of Pharmacy; and received an M.A. and Ph.D. degree from the University of Oklahoma. He served in the U.S. Army, was a Hospital Unit Training Officer during the Vietnam War, and is a retired Colonel of the Texas State Guard. He taught, inter alia, "History of Bioterrorism," at the Texas Military Academy, Camp Mabry, Austin, Texas. His biography appears in the current edition of Who's Who in America.

www.ingramcontent.com/pod-product-compliance
Lightning Source LLC
Chambersburg PA
CBHW030657270326
41929CB00007B/406